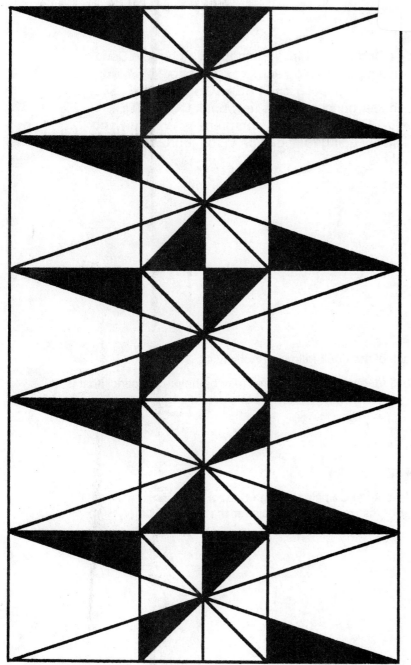

COMPUTE a DESIGN

WHOLE NUMBERS

Patricia Wright

JACOBS PUBLISHING COMPANY, INC.

PHOENIX   ARIZONA

AUTHOR   Patricia Wright

Patricia Wright is presently teaching at Thirkill Elementary School in Soda Springs, Idaho. She formerly taught at Dixon Junior High School in Provo, Utah. She is also the author of the popular book **SEARCH n SHADE.** She has applied the "search n shade" concept to this new book **COMPUTE a DESIGN: Whole Numbers** in order to provide a comprehensive review and practice program for students needing remediation in the fundamental operations with whole numbers.

COVER DESIGN   Robert E. Haberer

Robert E. Haberer is an art teacher at Camelback High School in Phoenix, Arizona. He was formerly the Director of Instructional Materials for the Phoenix Union High School District.

Printed in the United States of America

ISBN 0-918272-12-2

TABLE OF CONTENTS

1	ADDITION	Basic Facts; Three 1-Digit Numbers
2	ADDITION	Two 2-Digit Numbers, No Regrouping
3	ADDITION	Two 2-Digit Numbers, Regrouping 1's to 10's
4	ADDITION	Two 2-Digit Numbers, Regrouping 10's to 100's (some 1's to 10's)
5	ADDITION	Two 3-Digit Numbers, No Regrouping
6	ADDITION	Two 3-Digit Numbers, Regrouping 1's to 10's
7	ADDITION	Two 3-Digit Numbers, Regrouping 1's to 10's to 100's
8	ADDITION	Two 3-Digit Numbers, Regrouping 1's to 10's to 100's to 1000's
9	ADDITION	Three 2-Digit Numbers, Regrouping 1's to 10's to 100's
10	ADDITION	Three 3-Digit Numbers, Regrouping 1's to 10's to 100's to 1000's
11	ADDITION	Three or Four Numbers (at most 4 digits), Regrouping
12	REVIEW	Mix of Addition Exercises Based on Activities 1–11
13	SUBTRACTION	Basic Facts
14	SUBTRACTION	2-Digit Number Minus 1-Digit Number, No Regrouping
15	SUBTRACTION	2-Digit Number Minus 2-Digit Number, No Regrouping
16	SUBTRACTION	Two 3-Digit Numbers, No Regrouping
17	SUBTRACTION	2-Digit Number Minus 1-Digit Number, Regrouping
18	SUBTRACTION	2-Digit Number Minus 2-Digit Number, Regrouping
19	SUBTRACTION	3-Digit Number Minus 1-Digit Number, Regrouping 10's to 1's
20	SUBTRACTION	3-Digit Number Minus 2 or 3 Digit Number, Regrouping 10's to 1's
21	SUBTRACTION	3-Digit Number Minus 2-Digit Number, Regrouping 100's to 10's only
22	SUBTRACTION	3-Digit Number Minus 3-Digit Number, Regrouping 100's to 10's to 1's
23	SUBTRACTION	3-Digit Number (1 internal zero) Minus 3-Digit Number, Regrouping 100's to 10's to 1's
24	REVIEW	Mix of Addition and Subtraction Exercises
25	MULTIPLICATION	Basic Facts, Including Multiplication by 1 and 0
26	MULTIPLICATION	1-Digit Number × 2-Digit Number (some with regrouping 10's to 100's only)
27	MULTIPLICATION	1-Digit Number × 2-Digit Number, Regrouping
28	MULTIPLICATION	1-Digit Number × 3-Digit Number, Regrouping 1's to 10's (some 100's to 1000's)
29	MULTIPLICATION	1-Digit Number × 3-Digit Number (internal zero), Regrouping 1's to 10's (some 100's to 1000's)
30	MULTIPLICATION	1-Digit Number × 3-Digit Number, Regrouping 10's to 100's (some 100's to 1000's)

31	MULTIPLICATION	1-Digit Number × 3 or 4 Digit Number, Regrouping
32	MULTIPLICATION	2-Digit Number × 2-Digit Number (some with regrouping 10's to 100's to 1000's)
33	MULTIPLICATION	2-Digit Multiple of 10 × 2-Digit Multiple of 10 (some with regrouping 100's to 1000's)
34	MULTIPLICATION	2-Digit Number × 2-Digit Number, Regrouping
35	MULTIPLICATION	2-Digit Number × 3-Digit Number (some with internal zero), Regrouping
36	REVIEW	Mix of Multiplication Exercises Based on Activities 25–35
37	DIVISION	Basic Facts
38	DIVISION	1-Digit Divisor, 2-Digit Dividend, Nonzero Remainder
39	DIVISION	1-Digit Divisor, 2 or 3 Digit Dividend (every digit a multiple of divisor)
40	DIVISION	1-Digit Divisor, 2-Digit Dividend (first digit a multiple of divisor), Nonzero Remainder
41	DIVISION	1-Digit Divisor, 3-Digit Dividend (number formed by first two digits a multiple of divisor), Zero Remainder
42	DIVISION	1-Digit Divisor, 3 or 4 Digit Dividend (number formed by first two digits a multiple of divisor), Zero Remainder
43	DIVISION	1-Digit Divisor, 3-Digit Dividend, Zero and Nonzero Remainder
44	DIVISION	1-Digit Divisor, 3 or 4 Digit Dividend (quotient ends in zero), Nonzero Remainder
45	DIVISION	1-Digit Divisor, 4-Digit Dividend, Internal Zero in Quotient, Nonzero Remainder
46	DIVISION	2-Digit Divisor, 3-Digit Dividend, Zero Remainder
47	DIVISION	2-Digit Divisor, 4-Digit Dividend, 2-Digit Quotient, Zero Remainder
48	REVIEW	Mix of Division Exercises Based on Activities 37–47
	SOLUTION KEY	(See pages following Activity 48.)

Name_____

27	24	13	20	19	19	27	26	10	24
25	23	17	15	13	20	25	16	18	15
21	18	23	27	16	25	26	17	16	21
20	19	10	15	10	24	17	24	22	27
17	24	22	10	15	17	26	19	10	16
13	23	19	25	23	18	16	21	18	24
15	22	21	19	21	19	21	22	19	25
24	22	17	20	13	26	27	16	22	27
18	26	21	18	23	25	15	21	10	15
22	17	20	19	22	21	19	13	23	21

■
$$\begin{array}{r} 9 \\ 3 \\ +9 \\ \hline \end{array}$$

◩
$$\begin{array}{r} 7 \\ 9 \\ +8 \\ \hline \end{array}$$

◪
$$\begin{array}{r} 9 \\ +8 \\ \hline \end{array}$$

◪
$$\begin{array}{r} 9 \\ 9 \\ +9 \\ \hline \end{array}$$
◤
$$\begin{array}{r} 6 \\ 9 \\ +8 \\ \hline \end{array}$$
■
$$\begin{array}{r} 6 \\ 8 \\ +8 \\ \hline \end{array}$$
�
$$\begin{array}{r} 9 \\ 9 \\ +8 \\ \hline \end{array}$$

�￩
$$\begin{array}{r} 9 \\ 7 \\ +9 \\ \hline \end{array}$$
◣
$$\begin{array}{r} 6 \\ 6 \\ +6 \\ \hline \end{array}$$
◪
$$\begin{array}{r} 7 \\ +6 \\ \hline \end{array}$$
◪
$$\begin{array}{r} 8 \\ +8 \\ \hline \end{array}$$

◪
$$\begin{array}{r} 6 \\ +4 \\ \hline \end{array}$$
◪
$$\begin{array}{r} 7 \\ 3 \\ +5 \\ \hline \end{array}$$
■
$$\begin{array}{r} 8 \\ 5 \\ +6 \\ \hline \end{array}$$
◣
$$\begin{array}{r} 9 \\ 6 \\ +5 \\ \hline \end{array}$$

Name_____

86	26	99	85	86	26	96	55	89	26
86	89	96	75	89	86	69	85	89	86
69	47	26	21	25	47	92	86	99	47
92	73	55	25	57	21	75	25	92	21
26	86	69	92	86	26	73	55	26	89
86	89	21	47	26	89	99	92	86	26
75	25	57	21	55	25	85	21	75	25
21	85	26	69	96	57	75	86	96	92
86	26	99	92	89	26	73	75	26	86
26	89	73	75	26	86	69	85	26	86

■ 34
+52

◪ 43
+32

◩ 60
+13

◤ 57
+12

◤ 42
+15

■ 52
+37

◥ 15
+32

◣ 52
+44

◣ 11
+10

◤ 10
+15

◤ 62
+30

◤ 69
+30

◸ 54
+31

■ 14
+12

◥ 32
+23

3 ADDITION

Name_____

63	62	74	47	91	43	77	62	74	57
43	63	57	91	85	63	43	63	57	91
81	96	91	47	92	96	81	62	91	85
92	96	74	43	81	57	74	96	92	62
83	75	83	63	62	92	47	72	75	83
85	72	77	96	83	72	91	85	83	63
91	57	92	43	77	47	74	43	77	43
77	43	83	63	96	74	85	72	74	47
62	75	81	43	81	85	92	47	75	91
72	83	74	57	92	62	81	96	83	72

■ 36
+39

◥ 29
+18

◿ 24
+68

◸ 14
+49

◤ 36
+26

■ 35
+48

◥ 37
+48

◤ 57
+17

◣ 54
+37

◢ 58
+19

◸ 67
+29

◣ 42
+39

◸ 25
+18

■ 45
+27

◥ 29
+28

4 ADDITION

183	177	175	155	120	155	120	189	178	183
178	198	134	198	189	198	189	101	168	177
176	155	183	197	134	120	134	177	197	172
135	176	168	177	175	172	178	175	168	101
168	101	155	198	134	101	168	197	155	175
111	172	176	155	176	189	120	189	198	111
197	111	135	183	177	178	183	197	111	134
178	101	189	175	168	176	172	176	155	183
183	177	120	135	197	134	101	135	178	177
178	183	178	177	183	178	183	177	183	178

■ 83
+95

◧ 93
+96

◪ 75
+45

◨ 92
+83

◩ 73
+61

■ 96
+87

◪ 74
+94

◣ 99
+98

◤ 37
+64

◥ 99
+99

◧ 82
+73

◥ 98
+78

◨ 72
+63

■ 99
+78

◣ 91
+81

5 ADDITION

787	435	967	799	967	446	462	777	435	787
976	462	544	929	389	929	544	888	446	435
243	389	976	243	799	243	777	787	888	799
669	446	462	544	669	389	669	446	243	445
777	929	445	462	446	462	799	888	389	462
669	446	967	544	888	389	669	777	967	389
435	888	445	787	435	787	787	929	544	976
787	967	799	435	243	777	435	967	446	435
243	544	669	777	929	389	462	389	669	777
445	435	787	929	446	243	544	435	976	888

■ 624
+352

◣ 489
+310

◤ 811
+118

◢ 112
+131

◸ 423
+121

■ 123
+312

◺ 234
+212

◣ 452
+436

◤ 123
+546

◿ 245
+722

◹ 245
+144

◢ 341
+121

◿ 324
+121

■ 362
+425

◤ 541
+236

6 ADDITION

696	854	458	864	458	854	971	693	458	452
693	593	878	486	215	696	593	693	878	486
864	215	696	536	593	878	215	452	696	971
486	452	971	854	971	452	458	864	536	593
452	878	693	536	953	792	452	486	854	536
593	536	864	458	792	785	854	696	864	458
971	693	878	452	486	693	536	854	486	215
854	693	458	878	215	971	693	693	878	878
215	452	696	971	854	486	452	215	696	971
458	864	536	593	971	864	878	215	536	693

■   536
   +417

◣   127
   +325

◤   249
   +629

◹   254
   +717

◸   548
   +306

■   354
   +438

◺   107
   +108

◣   329
   +129

◤   239
   +247

◹   458
   +238

◸   237
   +356

◺   317
   +219

◸   436
   +257

■   569
   +216

◣   238
   +626

Name_____

662	724	826	901	662	833	826	901	724	833
833	647	888	563	724	662	600	888	616	662
803	964	600	616	833	724	647	761	908	593
761	826	616	733	593	803	563	647	593	600
647	563	908	593	908	964	826	761	733	616
600	901	647	761	662	724	733	901	803	563
593	733	964	833	803	901	662	908	964	826
908	901	724	647	563	733	616	833	803	964
662	733	593	600	593	803	761	647	563	833
833	724	600	616	908	964	826	761	724	662

■ 278
+384

◧ 734
+167

◪ 537
+196

◿ 659
+167

◸ 785
+179

■ 674
+159

◩ 278
+338

◣ 254
+346

◣ 729
+179

◹ 695
+108

◿ 493
+268

◿ 469
+178

◺ 368
+195

■ 267
+457

◩ 399
+194

8　ADDITION

1465	1332	1252	1074	1546	1332	1776	1513	1252	1334
1074	1930	1123	1371	1258	1377	1522	1377	1258	1546
1522	1776	1602	1334	1332	1776	1334	1465	1332	1930
1602	1258	1546	1513	1377	1258	1546	1074	1371	1465
1074	1252	1123	1371	1332	1776	1522	1377	1513	1546
1522	1377	1074	1252	1258	1930	1332	1776	1123	1371
1602	1513	1377	1123	1546	1074	1371	1522	1546	1465
1332	1930	1465	1602	1258	1930	1334	1465	1123	1252
1522	1252	1513	1546	1513	1776	1513	1776	1513	1930
1465	1258	1377	1123	1371	1258	1371	1522	1377	1602

■　676
　+789

◪　787
　+465

◩　528
　+994

◨　785　　◪　679　　■　809　　◪　659
　+289　　　+698　　　+793　　　+887

◣　734　　◣　579　　◪　696　　◨　874
　+389　　　+679　　　+817　　　+497

◪　469　　◨　943　　■　538　　◪　987
　+863　　　+987　　　+796　　　+789

152	175	209	173	207	212	149	175	209	152
189	152	186	152	219	139	186	215	152	200
175	188	215	200	149	173	188	186	158	209
219	207	186	173	189	219	207	152	175	188
212	139	152	219	209	212	189	215	219	149
200	209	152	173	188	158	149	186	212	139
173	139	215	158	207	175	188	215	158	207
158	149	186	212	139	219	209	186	173	189
207	215	152	186	175	149	186	215	152	212
152	200	189	219	188	158	139	200	189	215

■
```
  37
  64
+51
```

◪
```
  69
  84
+36
```

◣
```
  37
  56
+82
```

◿
```
  47
  88
+65
```

◨
```
  68
  92
+49
```

■
```
  98
  73
+44
```

◺
```
  77
  46
+65
```

◣
```
  54
  76
+43
```

◣
```
  98
  89
+25
```

◹
```
  98
  27
+94
```

◸
```
  79
  35
+93
```

◿
```
  34
  56
+68
```

◹
```
  54
  69
+26
```

■
```
  86
  23
+77
```

◺
```
  49
  39
+51
```

1232	1252	1323	2774	1987	1336	1734	2365	1987	1232
2774	1542	1252	1098	2365	1323	2365	1403	1622	1734
1098	2774	1542	1015	2218	1232	2218	1234	1987	1234
1336	1323	2218	1336	1403	1252	1987	1232	1622	1734
1622	1987	1232	1234	1098	2365	1323	1015	1336	1098
2774	1542	1015	2774	1734	1336	1403	1232	1234	1403
1234	1987	2218	1622	1542	1234	1098	1015	1252	1323
1403	2365	1734	2218	1015	1232	2218	1336	1098	2774
2365	1987	1622	1403	2774	1403	2774	1542	1252	1542
1015	1622	1734	1234	1098	1622	1323	1336	1323	1232

■
```
  849
  735
+ 634
```

◣
```
  676
  863
+ 448
```

◥
```
  965
  727
+ 673
```

◤
```
  232
  986
+  34
```

◸
```
  106
  943
+ 274
```

■
```
  147
  653
+ 215
```

�
```
  954
  459
+ 321
```

◣
```
  868
  735
+  19
```

◣
```
   34
  699
+ 501
```

◺
```
  873
  929
+ 972
```

◸
```
  578
  321
+ 643
```

◺
```
  932
  169
+ 235
```

◹
```
  287
  713
+  98
```

■
```
  392
  239
+ 601
```

◥
```
  459
  321
+ 623
```

11 ADDITION

3452	6435	8300	7697	9553	8169	4772	7004	3945	6435
6435	9553	8425	9553	7697	5694	7004	8323	8169	3945
9553	3761	4398	8425	4398	7004	8323	5736	5694	5736
3761	9553	3761	8300	3761	8323	8169	4772	8169	4772
4398	7697	9553	8425	3452	3945	4772	5736	5694	7004
8323	7004	5694	5736	6435	6435	8300	3761	8300	8425
8169	5694	8169	8323	7004	9553	7697	4398	7697	4398
8323	7004	4772	5736	4772	8425	8300	3761	8300	3761
6435	5694	5736	8323	8169	9553	8425	4398	8425	3945
3945	3452	4772	8169	5694	7697	9553	7697	6435	6435

■
```
  3242
   326
  2653
+  214
```

◣
```
  1289
  2480
   394
 +4006
```

◥
```
  2527
   354
   231
 +1660
```

◿
```
  8969
   494
    75
 +  15
```

◸
```
   697
   930
  6058
 +  12
```

■
```
  2969
   132
 + 844
```

◥
```
  6214
   697
    12
 +  81
```

◣
```
  4126
   964
  3221
 +  12
```

◣
```
   158
  4643
   851
 +  42
```

◿
```
  1494
   153
  1277
 +1474
```

◿
```
   424
   932
   241
 +2164
```

◿
```
  7698
    57
   494
 +  51
```

◸
```
  7008
   994
   321
 + 102
```

■
```
    46
   978
 +2428
```

◥
```
  2268
  1409
   355
 +1704
```

Name_____

16	19	112	36	19	8	112	48	16	19
8	134	1009	142	77	112	1432	579	36	8
19	1432	122	77	579	1009	122	48	1212	16
16	48	1212	123	122	77	142	1432	134	19
8	1212	36	122	123	1212	36	134	123	8
36	19	579	123	16	8	579	1009	16	134
123	16	77	112	48	134	36	134	19	142
36	8	142	1432	142	1009	579	1432	8	112
579	77	19	122	77	112	36	16	134	1009
19	1212	48	1212	1009	1212	1432	122	123	8

■
$$\begin{array}{r} 5 \\ 6 \\ +5 \end{array}$$

◪
$$\begin{array}{r} 15 \\ +21 \end{array}$$

◺
$$\begin{array}{r} 49 \\ +93 \end{array}$$

◿
$$\begin{array}{r} 47 \\ +65 \end{array}$$

◹
$$\begin{array}{r} 24 \\ 34 \\ +65 \end{array}$$

■
$$\begin{array}{r} 6 \\ 9 \\ +4 \end{array}$$

◤
$$\begin{array}{r} 45 \\ +32 \end{array}$$

◣
$$\begin{array}{r} 369 \\ +843 \end{array}$$

◣
$$\begin{array}{r} 132 \\ +447 \end{array}$$

◿
$$\begin{array}{r} 88 \\ +46 \end{array}$$

◹
$$\begin{array}{r} 356 \\ +653 \end{array}$$

◿
$$\begin{array}{r} 76 \\ +46 \end{array}$$

◸
$$\begin{array}{r} 364 \\ 645 \\ +423 \end{array}$$

■
$$\begin{array}{r} 3 \\ 4 \\ +1 \end{array}$$

◤
$$\begin{array}{r} 11 \\ +37 \end{array}$$

13 SUBTRACTION

Name _____

7	14	10	11	2	4	8	5	8	14
12	5	8	9	6	4	16	15	10	3
15	1	3	2	1	15	2	12	15	1
16	5	6	7	9	10	14	6	16	3
6	2	8	5	7	15	16	11	4	2
4	6	12	11	10	5	1	3	6	2
8	14	4	16	14	1	9	4	7	15
9	12	15	6	10	3	2	8	9	12
7	11	16	11	2	6	1	5	7	11
10	9	8	3	6	4	16	11	12	3

■ $\begin{array}{r} 5 \\ -3 \\ \hline \end{array}$

◪ $\begin{array}{r} 13 \\ -\ 2 \\ \hline \end{array}$

◪ $\begin{array}{r} 19 \\ -\ 3 \\ \hline \end{array}$

◸ $\begin{array}{r} 9 \\ -2 \\ \hline \end{array}$ ◤ $\begin{array}{r} 9 \\ -6 \\ \hline \end{array}$ ■ $\begin{array}{r} 6 \\ -2 \\ \hline \end{array}$ ◪ $\begin{array}{r} 15 \\ -\ 1 \\ \hline \end{array}$

◣ $\begin{array}{r} 12 \\ -\ 2 \\ \hline \end{array}$ ◣ $\begin{array}{r} 16 \\ -\ 4 \\ \hline \end{array}$ ◸ $\begin{array}{r} 8 \\ -0 \\ \hline \end{array}$ ◸ $\begin{array}{r} 6 \\ -1 \\ \hline \end{array}$

◣ $\begin{array}{r} 6 \\ -5 \\ \hline \end{array}$ ◸ $\begin{array}{r} 10 \\ -\ 1 \\ \hline \end{array}$ ■ $\begin{array}{r} 7 \\ -1 \\ \hline \end{array}$ ◪ $\begin{array}{r} 19 \\ -\ 4 \\ \hline \end{array}$

14 SUBTRACTION

97	64	82	97	85	24	82	97	64	82
82	85	42	91	60	34	42	85	70	97
64	34	64	55	82	64	60	97	32	64
97	91	70	97	64	97	64	63	70	82
63	32	64	82	63	24	97	82	96	42
96	24	82	97	55	45	64	97	91	32
64	34	45	64	82	64	82	34	60	97
97	85	97	63	97	82	70	97	70	82
82	55	32	34	42	91	32	34	45	64
64	97	64	82	96	60	97	82	64	97

■
$$89 \\ -\ 7$$

◪
$$48 \\ -\ 6$$

◥
$$99 \\ -\ 3$$

◿
$$89 \\ -\ 4$$

◸
$$46 \\ -\ 1$$

■
$$67 \\ -\ 3$$

◹
$$75 \\ -\ 5$$

◩
$$59 \\ -\ 4$$

◣
$$36 \\ -\ 2$$

◤
$$97 \\ -\ 6$$

◿
$$38 \\ -\ 6$$

◿
$$67 \\ -\ 4$$

◸
$$63 \\ -\ 3$$

■
$$99 \\ -\ 2$$

◥
$$25 \\ -\ 1$$

15 SUBTRACTION

Name _____

61	33	61	10	20	33	43	10	20	11
8	26	13	36	2	26	33	14	2	36
43	8	24	17	14	13	12	24	20	11
20	11	17	12	43	11	24	17	26	13
2	36	10	26	8	20	2	43	8	61
20	11	14	13	61	10	36	33	26	8
24	36	10	43	8	14	11	43	13	12
17	26	33	20	10	61	8	14	10	17
43	8	43	2	36	33	43	2	36	11
13	61	13	14	11	26	13	14	10	20

■ 68
−44

◪ 86
−75

◩ 32
−12

◨ 98
−37

◪ 66
−33

■ 84
−72

◩ 89... 48
−38

◪ 59
−23

◪ 65
−51

◨ 89
−46

◨ 98
−85

◨ 69
−43

◨ 99
−91

■ 38
−21

◪ 87
−85

16 SUBTRACTION

200	612	100	311	700	132	323	612	231	311
777	323	212	116	212	200	777	223	212	116
521	231	241	700	100	241	700	231	777	521
125	521	132	323	132	700	125	132	521	323
811	311	223	777	521	132	116	212	200	811
231	811	612	323	777	116	311	100	811	241
700	100	212	100	125	223	241	323	777	700
132	521	116	125	700	521	200	612	521	132
521	521	132	231	811	811	241	700	132	521
700	132	700	132	521	700	132	521	132	700

■ 900
−200

◣ 534
−223

◤ 229
−113

◹ 600
−400

◸ 677
−436

■ 346
−214

◥ 435
−223

◣ 643
−412

◣ 700
−600

◹ 459
−236

◸ 922
−310

◤ 954
−631

◤ 999
−222

■ 928
−407

◥ 829
−704

17 SUBTRACTION

59	48	59	69	26	37	46	59	88	48
48	88	77	18	59	48	19	53	88	59
59	76	56	59	96	64	48	28	84	48
84	28	53	88	95	67	59	77	18	77
28	53	19	46	48	59	76	56	76	26
88	37	46	37	53	69	18	69	18	59
48	59	19	84	28	26	76	56	88	48
46	88	48	28	53	69	26	48	59	77
37	84	59	88	19	18	59	88	69	56
88	19	53	48	59	88	48	77	26	59

■
$$64 - 5$$

�
$$50 - 4$$

◣
$$42 - 5$$

◿
$$78 - 9$$

◺
$$33 - 7$$

■
$$95 - 7$$

◥
$$61 - 8$$

◤
$$22 - 3$$

◤
$$36 - 8$$

◸
$$82 - 6$$

◿
$$63 - 7$$

◿
$$84 - 7$$

◸
$$27 - 9$$

■
$$57 - 9$$

◥
$$92 - 8$$

Name_____

69	67	75	37	73	73	75	26	73	67
73	9	27	25	26	68	58	25	8	69
68	58	67	73	22	32	67	73	6	37
6	8	69	67	9	8	73	69	9	32
69	25	37	75	32	6	37	68	58	73
73	68	27	22	26	75	32	22	26	69
9	32	69	67	6	27	69	67	25	8
25	37	73	73	9	8	67	73	75	27
73	22	26	68	27	22	37	9	32	67
67	69	6	58	69	67	6	58	69	73

■ 90
  −17

◨ 64
  −38

◪ 64
  −39

◸ 92   ◹ 85   ■ 82   ◳ 35
  −17   −27   −15   −27

◣ 22   ◤ 91   ◹ 64   ◳ 81
  −16   −69   −55   −49

◸ 93   ◰ 62   ■ 77   ◳ 86
  −25   −35   − 8   −49

389	247	927	128	667	734	389	189	927	128
722	239	516	628	444	128	722	239	417	628
734	444	128	667	369	247	464	389	444	516
417	369	189	464	444	128	667	369	722	464
927	516	239	247	369	722	628	516	239	189
734	128	722	734	417	734	516	927	444	667
389	189	239	444	722	628	128	667	369	128
128	667	628	247	464	516	369	189	239	389
369	444	516	239	128	444	417	734	128	722
417	628	128	389	189	927	128	389	247	464

■ 391
  − 2

◨ 936
  − 9

◪ 524
  − 8

◸ 254
 − 7

◹ 742
 − 8

■ 137
 − 9

◺ 637
 − 9

◣ 673
 − 6

◣ 425
 − 8

◺ 193
 − 4

◹ 472
 − 8

◺ 731
 − 9

◸ 247
 − 8

■ 452
 − 8

◺ 376
 − 7

Name_____

419	259	617	419	259	617	419	259	617	419
617	717	358	606	827	338	606	349	309	617
259	546	527	259	438	828	419	438	228	259
419	606	419	617	259	617	259	617	606	419
617	459	228	259	338	228	617	358	717	617
259	338	527	617	459	828	419	459	827	259
419	606	617	419	617	419	259	419	606	419
617	438	827	259	338	228	617	546	527	617
259	349	309	606	828	459	606	717	546	259
419	259	617	419	259	617	419	259	617	419

■ 456
  − 37

◤ 874
  −525

◣ 321
  − 12

◹ 685
  −327

◸ 986
  −459

■ 654
  − 37

◥ 247
  − 19

◤ 568
  −109

◤ 673
  −235

◹ 965
  −627

◸ 745
  − 28

◹ 593
  − 47

◸ 854
  − 26

■ 972
  −713

◥ 865
  − 38

53	795	452	53	162	53	791	795	86	162
880	91	862	783	791	162	63	91	493	783
493	86	53	675	86	880	195	53	63	91
162	675	783	162	493	232	162	880	232	162
791	53	862	452	791	791	795	195	791	53
162	791	63	232	53	53	862	452	53	162
53	880	195	162	63	452	162	493	783	791
795	91	791	795	232	493	86	791	675	86
862	86	880	91	791	162	675	783	63	91
791	675	195	53	162	53	791	493	232	53

■   137
      − 84

◨   514
      − 62

◨   925
      − 63

◸   137
     − 74

◹   188
     − 97

■   258
     − 96

�￬   864
     − 81

�￬   529
     − 36

◤   757
     − 82

◿   962
     − 82

◹   238
     − 43

◿   867
     − 72

◺   316
     − 84

■   842
     − 51

◤   137
     − 51

Name_____

366	257	369	686	488	286	585	257	369	277
677	386	179	286	585	277	677	386	683	588
277	339	588	366	683	179	686	488	587	366
386	686	369	386	683	386	683	257	369	179
683	587	339	257	366	686	369	286	339	386
587	369	686	488	587	339	588	585	686	488
366	179	386	257	585	257	369	683	179	277
677	386	683	386	286	677	386	179	386	587
686	488	386	179	257	585	683	683	588	366
179	277	339	683	386	179	386	286	369	683

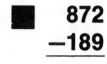
■ 872
−189

◥ 963
−286

◣ 453
−196

◸ 435
−149

◸ 565
−199

■ 423
−244

◥ 776
−288

◣ 976
−699

◣ 954
−268

◿ 941
−354

◸ 881
−296

◿ 917
−329

◿ 768
−399

■ 923
−537

◥ 828
−489

23 SUBTRACTION

Name_____

769	638	468	769	638	468	166	568	638	769
638	769	468	39	38	239	279	137	38	638
468	638	239	368	174	215	39	466	174	466
769	166	215	769	638	769	335	279	39	368
468	137	568	468	239	38	468	166	279	769
638	166	279	769	174	215	769	335	38	769
39	368	166	568	468	468	638	239	279	638
335	568	137	368	239	466	39	215	638	468
769	174	38	166	279	174	368	468	468	769
638	468	137	215	769	638	468	638	769	468

■ 907 −138

◨ 407 −369

◧ 503 −168

◸ 402 −163   ◤ 704 −425   ■ 803 −335   ◺ 904 −336

◧ 305 −168   ◩ 301 −127   ◿ 203 −164   ◣ 706 −338

◿ 605 −439   ◪ 602 −387   ■ 907 −269   ◨ 903 −437

77	231	8	22	761	185	22	28	25	427
477	22	28	15	477	25	84	8	22	16
84	8	8	67	427	761	231	28	8	15
567	22	28	77	567	16	427	22	28	25
22	8	761	231	77	185	67	185	8	22
28	22	16	185	25	477	15	477	28	8
427	8	28	25	84	761	567	28	22	761
477	22	8	761	567	16	84	8	28	67
84	28	22	25	84	77	477	22	8	77
67	185	8	22	16	231	8	28	15	231

■
46
−18

◥
774
−347

◣
5
3
+8

◸
90
−13

◿
467
−236

■
16
− 8

◥
304
−119

◣
53
−28

◥
86
−19

◸
9
+6

◿
428
+139

◸
824
− 63

◿
764
−287

■
54
−32

◥
90
− 6

Name_____

45	16	9	63	0	45	25	12	45	0
0	45	16	18	81	30	18	0	16	45
54	0	45	12	16	0	9	45	0	36
30	18	35	45	3	63	45	72	18	35
0	81	16	36	12	30	64	16	30	16
45	64	45	72	64	36	12	45	3	45
36	18	54	16	9	81	45	36	18	64
12	45	16	63	45	16	25	0	45	9
45	0	16	18	64	3	18	16	0	16
16	45	25	81	16	45	72	54	45	45

5
×5

◣ 3
×3

◹ 1
×3

◹ 9   ◼ 4   ◼ 9   ◪ 9
×6   ×4   ×5   ×9

◤ 8 × 9 = ___   ◹ 6   ◣ 7   ◣ 5
                    ×6   ×9   ×6

 5 × 7 = ___   ◤ 8 × 8 = ___   ◼ 0 × 9 = ___   3
                                                                        ×4

26 MULTIPLICATION

159	213	0	96	159	168	148	637	213	168
168	120	86	49	426	120	288	49	96	159
148	288	0	637	180	189	148	96	129	426
180	96	129	195	637	0	186	189	148	86
213	129	426	49	231	195	86	0	288	213
159	0	288	0	186	184	426	49	637	159
120	86	120	192	189	129	198	637	180	426
49	96	180	288	120	426	49	288	120	189
168	180	637	148	189	49	96	148	288	159
213	159	129	288	168	213	129	86	168	213

71
× 3

91
× 7

90
× 2

78 × 0	43 × 2	53 × 3	32 × 3
49 × 1	43 × 3	74 × 2	63 × 3
40 × 3	72 × 4	84 × 2	71 × 6

27 MULTIPLICATION

Name ____________________

168	372	352	416	256	344	352	324	136	335
462	335	180	136	168	335	344	256	168	456
372	295	456	295	252	168	416	462	324	256
416	350	180	942	335	252	942	350	180	352
256	168	335	252	168	335	252	168	335	372
462	252	335	168	252	168	335	252	168	324
180	352	324	942	168	335	942	462	456	136
324	350	180	136	335	252	344	256	180	295
136	252	416	295	252	168	416	295	252	372
335	456	350	344	462	324	350	180	352	168

■ 24 × 7

◥ 59 × 5

◣ 43 × 8

◤ 52 × 8

◢ 32 × 8

■ 67 × 5

◥ 66 × 7

◣ 62 × 6

◣ 30 × 6

◤ 76 × 6

◢ 68 × 2

◤ 36 × 9

◢ 70 × 5

■ 84 × 3

◥ 88 × 4

1278	942	436	6496	657	1878	987	672	942	1278
1652	249	858	460	6496	987	858	657	249	942
1251	436	249	972	1278	942	1578	249	972	460
672	1256	672	1562	858	1256	1521	987	1251	436
460	6496	1652	987	531	485	672	942	436	858
987	858	942	1256	627	924	858	1652	460	672
1878	436	1251	1587	972	987	6469	1256	972	1256
6496	657	249	1878	942	1652	460	249	1251	987
1278	249	972	987	858	657	6496	987	249	942
942	1278	460	1251	1578	672	1256	858	942	1652

■ 213
× 6

◩ 626
× 3

◫ 218
× 2

◪ 219
× 3

◩ 812
× 8

■ 314
× 3

◳ 429
× 2

◤ 329
× 3

◤ 789
× 2

◥ 314
× 4

◥ 324
× 3

◥ 115
× 4

◰ 336
× 2

■ 413
× 4

◣ 417
× 3

Name_____

2114	982	624	2114	918	520	624	2114	642	2114
642	2141	642	728	624	2114	728	642	782	642
728	782	2114	6318	1421	1025	1212	2114	642	2114
2114	624	7218	327	410	3521	918	5472	728	624
1212	2114	1218	5472	624	728	6318	520	2114	1025
327	728	1025	3521	728	2114	1218	1421	624	410
624	2114	918	1421	7218	5472	1025	3521	728	2114
728	642	2114	410	520	1218	327	2114	2141	624
2141	782	642	728	624	728	624	782	642	782
624	2141	728	2114	6318	5472	2114	624	782	2114

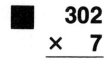

$$302 \times 7$$

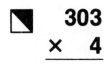

$$303 \times 4$$

◩ $$306 \times 3$$

◩ $$702 \times 9$$

◩ $$503 \times 7$$

■ $$104 \times 7$$

◩ $$608 \times 9$$

◪ $$203 \times 6$$

◪ $$205 \times 2$$

◳ $$205 \times 5$$

◲ $$104 \times 5$$

◳ $$802 \times 9$$

◱ $$109 \times 3$$

■ $$104 \times 6$$

◩ $$203 \times 7$$

Name_____

306	2286	5220	1056	504	1740	2346	306	2286	5220
2286	5220	1740	1720	768	2380	1928	504	5220	306
2346	306	768	1449	1740	504	1056	1155	2286	1768
1928	504	2286	2328	1155	2328	1720	5220	1740	1155
5220	2328	1449	5220	1056	2346	306	1056	1720	306
306	1768	2380	306	768	1720	5220	2328	1449	2286
1740	1720	2286	1768	1449	1768	1449	2286	1928	504
2380	5220	1056	1155	2328	1720	768	504	306	2328
306	2286	768	2346	1056	504	1740	2380	5220	306
5220	306	5220	1928	1720	768	1155	306	2286	5220

153
× 2

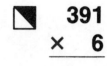
391
× 6

384
× 2

352   231   381   252
× 3   × 5   × 6   × 2

241   582   442   430
× 8   × 4   × 4   × 4

290   340   580   483
× 6   × 7   × 9   × 3

Name_____

10936	1395	6265	1645	1395	1692	2352	2844	1824	1645
1368	3356	1692	2352	7950	1368	1645	9464	1167	7950
9464	1338	10936	2844	1824	9464	6265	10936	1395	1824
1645	2352	3356	1338	2352	1645	1395	1167	2352	1645
1368	7950	1692	10936	2844	6265	10936	2170	1368	3356
1824	1395	6265	2352	1395	1824	2352	2844	1338	2170
1645	2352	2170	1167	1645	2352	7950	1692	1645	2352
2844	1167	10936	9464	1368	3356	1824	2352	3356	1167
1692	9464	1368	1645	1395	1338	10936	7950	1338	1395
10936	7950	1692	2352	2844	1167	1645	2170	6265	2352

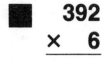
$$392 \times 6$$

$$456 \times 3$$

$$434 \times 5$$

$$839 \times 4$$ $$669 \times 2$$ $$235 \times 7$$ $$1253 \times 5$$

$$1352 \times 7$$ $$465 \times 3$$ $$948 \times 3$$ $$456 \times 4$$

$$1325 \times 6$$ $$564 \times 3$$ $$2734 \times 4$$ $$389 \times 3$$

Name_____

756	253	483	1606	736	1008	1749	308	253	989
989	682	528	756	253	989	756	852	2046	756
714	1008	989	714	837	483	2046	253	736	837
1606	253	682	528	852	1606	736	308	989	1749
756	989	834	2046	714	308	483	834	253	756
308	714	1008	1749	834	438	528	1749	837	714
852	1606	989	483	1606	1749	837	253	852	1606
989	756	682	528	989	756	852	2046	756	253
253	989	736	308	756	253	483	1008	253	989
438	834	438	1008	989	756	736	438	438	834

■ 23
×11

◧ 22
×14

◪ 53
×33

◹ 21
×23

◩ 73
×22

■ 43
×23

◸ 62
×33

◺ 71
×12

◺ 32
×23

◹ 34
×21

◹ 42
×24

◹ 22
×31

◿ 44
×12

■ 21
×36

◺ 31
×27

33 MULTIPLICATION

Name_____

200	400	2000	4200	800	300	6300	4200	800	500
400	3000	1400	1800	4200	800	1800	1400	3000	6300
4500	1400	1800	300	900	1600	800	1800	1400	2500
400	2400	300	2800	4200	2000	2800	6300	2400	2000
4500	400	1200	4200	300	800	2000	900	6300	3600
400	100	1600	1500	4500	3600	1000	1200	1000	2000
1500	2400	100	2800	4500	2500	2800	3600	2400	1000
400	1400	1800	4500	1600	1200	1000	1800	1400	800
100	3000	1400	1800	1500	1000	1800	1400	3000	3600
500	100	3600	1500	2500	4500	1000	100	2500	200

30
×30

50
×90

50
×40

50
×50

20
×20

40
×40

10
×10

80
×10

90
×70

50
×20

60
×70

60
×60

30
×10

30
×40

30
×50

34 MULTIPLICATION

Name_____

1230	4141	4011	1230	4141	2250	4928	2210	4928	4187
4011	4187	4928	1125	2340	2210	6256	4982	4178	2210
4141	4928	2210	2106	2184	4187	2345	2201	4982	4187
1230	2345	2184	4928	6045	2646	6045	2340	2106	4928
4011	1566	2106	2646	4187	6256	2652	2210	4928	2052
2250	4187	2210	6256	2184	2210	2106	2646	2052	1230
4928	2646	2052	2184	6045	2646	4187	6045	2340	4141
2210	4982	2201	1566	2210	2106	2652	4928	4187	4011
4928	4178	4982	2652	4928	1125	2250	2210	4928	1230
4187	2210	4187	4928	2345	1230	4141	1230	4011	4141

■ 79 ×53

◨ 56 ×39

◪ 68 ×92

◩ 67 ×35   ◪ 39 ×60   ■ 65 ×34   ◪ 49 ×54

◪ 54 ×39   ◪ 93 ×65   ◩ 54 ×38   ◪ 29 ×54

◩ 45 ×25   ◪ 45 ×50   ■ 88 ×56   ◪ 78 ×34

Name_____

15330	37668	23460	37668	13685	19980	33885	49082	36426	34224
14196	18496	11418	15330	11418	33885	24360	23532	49082	13685
13685	14196	23460	30618	18496	34224	33885	34224	14196	19980
19980	13685	23532	34224	14196	13685	15330	30618	13685	14196
14196	33885	49082	13685	36426	30618	19980	23460	11418	19980
13685	15330	37668	14196	15330	24360	14196	36426	49082	13685
19980	13685	18496	11418	14196	19980	33885	24360	19980	14196
14196	19980	23532	24360	23532	30618	23460	30618	13685	19980
19980	33885	49082	33885	49082	18496	11418	15330	37668	13685
36426	34224	36426	34224	14196	13685	15330	37668	18496	30618

595
× 23

876
× 43

345
× 68

753
× 45

506
× 97

555
× 36

346
× 33

578
× 32

365
× 42

467
× 78

368
× 93

444
× 53

435
× 56

507
× 28

567
× 54

Name_____

72	56	810	56	72	810	56	72	56	810
810	30	850	330	42	636	1020	330	850	56
56	280	124	100	64	9200	306	100	124	72
810	30	42	810	72	56	810	636	850	56
330	306	100	850	56	72	330	124	9200	42
280	850	330	124	810	56	9200	42	636	124
56	100	64	56	72	810	72	100	306	810
810	636	42	636	1020	30	850	30	1020	72
72	9200	64	9200	124	280	306	280	124	56
56	810	72	56	810	810	72	56	72	810

45
×18

34
×25

◪ **100**
× 92

◩ **5**
×6

◪ **16**
× 4

◼ **36**
× 2

◺ **7**
×6

◪ **56**
× 5

◪ **25**
× 4

◪ **106**
× 6

◪ **102**
× 3

◪ **33**
×10

◩ **31**
× 4

◼ **14**
× 4

◪ **34**
×30

Name_____

4	1	5	7	0	8	7	6	5	4
2	12	9	10	5	6	1	3	12	10
11	5	7	0	3	8	11	4	2	0
12	8	3	4	10	1	7	9	3	12
0	9	0	9	12	7	3	11	0	11
10	6	5	1	12	4	5	2	10	1
7	6	10	4	0	8	4	2	10	7
2	3	12	5	5	6	1	12	8	10
9	7	6	3	0	11	8	5	7	3
4	11	3	7	10	2	7	9	0	12

◺ $3\overline{)33}$

◹ $11\overline{)55}$ ◺ $7\overline{)63}$ ■ $9\overline{)36}$ ◿ $12\overline{)24}$

◸ $9\overline{)27}$ ◺ $7\overline{)56}$ ◹ $8\overline{)8}$ ■ $5\overline{)60}$

◿ $12\overline{)72}$ ◺ $11\overline{)0}$ ■ $9\overline{)63}$ ◺ $4\overline{)40}$

Name_____

9 r1	4 r3	4 r2	9 r1	5 r1	3 r5	9 r1	3 r1	4 r3	9 r1
2 r4	3 r6	3 r1	8 r3	3 r4	2 r7	6 r3	9 r1	2 r4	3 r6
4 r2	3 r1	6 r2	3 r1	2 r8	6 r2	4 r2	2 r8	9 r1	4 r2
9 r1	5 r1	3 r1	4 r2	6 r1	5 r4	4 r2	3 r1	6 r3	9 r1
8 r3	6 r1	2 r7	5 r3	5 r1	3 r5	3 r8	3 r4	3 r8	3 r5
2 r7	2 r8	5 r1	6 r3	2 r7	5 r3	5 r1	6 r3	6 r2	5 r3
9 r1	3 r8	4 r2	3 r1	3 r5	8 r3	3 r1	4 r2	6 r1	9 r1
3 r1	9 r1	5 r4	4 r2	6 r1	5 r4	3 r1	3 r4	4 r2	3 r1
3 r6	2 r4	3 r1	2 r7	2 r8	8 r3	5 r3	4 r2	3 r6	2 r4
9 r1	4 r3	4 r2	9 r1	3 r8	6 r1	9 r1	4 r2	4 r3	9 r1

■ 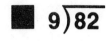 $9\overline{)82}$

◳ $7\overline{)26}$

◨ $5\overline{)29}$

◲ $3\overline{)16}$ ◪ $5\overline{)19}$ ■ $3\overline{)10}$ ◳ $6\overline{)39}$

◨ $8\overline{)23}$ ◨ $9\overline{)35}$ ◱ $3\overline{)20}$ ◱ $4\overline{)23}$

◱ $9\overline{)75}$ ◱ $2\overline{)13}$ ■ $4\overline{)18}$ ◨ $9\overline{)26}$

Name _____

10	23	12	101	23	111	423	11	111	10
111	212	121	43	322	212	133	321	121	23
121	21	133	23	21	220	111	43	220	321
43	101	10	321	11	43	322	10	212	133
10	12	322	133	423	101	12	423	11	111
111	423	220	121	12	133	212	21	121	23
321	11	23	21	101	321	220	111	43	101
133	212	322	10	423	322	10	423	322	12
23	12	11	321	11	43	121	43	220	10
111	10	212	133	111	10	21	101	111	23

■ 8)‾888‾

◩ 3)‾966‾

◪ 4)‾84‾

◸ 4)‾848‾   ◹ 8)‾88‾   ■ 7)‾70‾   ◩ 4)‾484‾

◩ 4)‾48‾   ◩ 2)‾86‾   ◿ 2)‾846‾   ◺ 3)‾660‾

◸ 3)‾963‾   ◺ 3)‾399‾   ■ 3)‾69‾   ◩ 5)‾505‾

44 r1	22 r1	22 r2	11 r1	22 r1	12 r1	10 r5	10 r8	12 r1	44 r1
12 r1	44 r1	32 r2	10 r4	21 r3	10 r8	10 r3	31 r1	22 r1	22 r1
10 r3	10 r4	22 r1	32 r1	31 r1	32 r1	31 r1	44 r1	22 r2	10 r8
11 r2	32 r1	10 r4	12 r1	22 r1	44 r1	12 r1	21 r3	31 r1	32 r2
12 r1	21 r3	11 r1	44 r1	10 r3	10 r8	22 r1	32 r2	20 r1	44 r1
44 r1	10 r5	10 r8	22 r1	10 r5	11 r1	44 r1	22 r2	11 r1	22 r1
10 r8	22 r2	31 r1	12 r1	44 r1	22 r1	12 r1	32 r2	10 r4	21 r3
10 r5	11 r2	22 r1	21 r3	10 r8	10 r3	10 r4	22 r1	32 r1	11 r1
22 r1	44 r1	22 r2	11 r2	32 r1	11 r2	10 r5	20 r1	44 r1	12 r1
12 r1	44 r1	32 r1	20 r1	22 r1	12 r1	21 r3	11 r1	22 r1	44 r1

■ $3\overline{)37}$

�￩ $6\overline{)64}$

◥ $3\overline{)98}$

◤ $3\overline{)68}$ ◣ $3\overline{)94}$ ■ $2\overline{)89}$ ◥ $3\overline{)61}$

◥ $6\overline{)65}$ ◥ $2\overline{)65}$ ◤ $7\overline{)73}$ ◢ $7\overline{)79}$

◤ $4\overline{)87}$ ◢ $8\overline{)89}$ ■ $3\overline{)67}$ ◥ $9\overline{)98}$

Name _____

71	52	91	82	52	42	41	32	71	52
42	71	18	18	51	63	18	18	52	42
92	18	52	71	31	61	42	71	18	21
91	18	42	41	82	63	51	52	18	62
52	81	82	31	32	81	62	63	61	71
71	63	62	21	82	21	51	31	82	42
21	18	71	91	61	31	32	71	18	92
32	18	42	52	63	92	52	42	18	31
42	52	18	18	61	91	18	18	42	71
71	42	21	62	52	42	81	51	71	52

$6\overline{)426}$

$3\overline{)246}$

$9\overline{)819}$

$2\overline{)126}$ $4\overline{)128}$ $3\overline{)126}$ $6\overline{)306}$

$5\overline{)155}$ $8\overline{)648}$ $5\overline{)105}$ $6\overline{)366}$

$4\overline{)164}$ $3\overline{)186}$ $3\overline{)156}$ $3\overline{)276}$

914	848	31	90	510	317	914	848	510	90
163	612	836	816	836	414	410	163	612	410
31	410	31	848	411	512	914	848	816	317
414	317	816	623	80	512	414	623	914	623
510	623	512	512	510	90	512	512	414	90
163	90	411	80	163	410	80	411	31	836
914	410	914	317	512	512	31	848	163	848
414	848	816	836	411	411	163	836	510	623
510	612	317	510	90	914	317	31	612	90
816	836	163	410	816	623	414	836	816	410

■ $7\overline{)2877}$

◥ $2\overline{)1696}$

◤ $3\overline{)489}$

◪ $2\overline{)1828}$ ◪ $2\overline{)1672}$ ■ $8\overline{)640}$ ◥ $9\overline{)810}$

◣ $6\overline{)4896}$ ◣ $6\overline{)2484}$ ◪ $5\overline{)2550}$ ◪ $9\overline{)3690}$

◪ $8\overline{)248}$ ◪ $3\overline{)1869}$ ■ $3\overline{)1536}$ ◥ $5\overline{)1585}$

43 DIVISION

Name_____

122 r2	362 r1	63	53	13	67	66	53	62	362 r1
62	66	25	112	42 r1	54 r4	58	66 r2	67	62
13	58	66	53	122 r2	62	63	67	54 r4	328 r2
112	328 r2	66 r2	58	66	53	66 r2	42 r1	63	58
63	42 r1	362 r1	63	25	54 r4	67	362 r1	66 r2	67
66 r2	67	62	42 r1	62	122 r2	66 r2	62	66	58
66	58	122 r2	63	328 r2	13	67	122 r2	112	53
54 r4	53	362 r1	122 r2	62	362 r1	62	362 r1	13	25
362 r1	66 r2	328 r2	13	67	63	53	66	58	62
62	122 r2	54 r4	25	54 r4	58	66 r2	42 r1	362 r1	122 r2

■ 3⟌368

◹ 4⟌212

◣ 5⟌560

◸ 7⟌441   ◨ 7⟌175   ■ 9⟌558   ◺ 9⟌603

◣ 5⟌274   ◣ 6⟌398   ◸ 8⟌104   ◨ 8⟌337

◸ 6⟌396   ◪ 7⟌406   ■ 2⟌725   ◺ 3⟌986

44 DIVISION

20 r1	230 r3	270 r2	660	760 r1	940 r3	270 r2	220 r3	230 r3	20 r1
330 r4	230 r3	690 r6	240 r2	270 r2	440 r2	810 r2	120 r4	330 r4	230 r3
190 r3	660	190 r3	220 r3	760 r1	940 r3	270 r2	220 r3	70 r2	940 r3
810 r2	240 r2	690 r6	120 r4	190 r3	240 r2	810 r2	120 r4	760 r1	240 r2
660	70 r2	220 r3	70 r2	440 r2	20 r1	70 r2	660	70 r2	940 r3
810 r2	440 r2	690 r6	120 r4	230 r3	270 r2	240 r2	690 r6	240 r2	810 r2
190 r3	660	70 r2	940 r3	190 r3	120 r4	190 r3	220 r3	270 r2	220 r3
810 r2	440 r2	690 r6	440 r2	760 r1	660	690 r6	440 r2	760 r1	120 r4
330 r4	20 r1	270 r2	220 r3	190 r3	440 r2	70 r2	660	20 r1	330 r4
230 r3	20 r1	760 r1	240 r2	690 r6	940 r3	810 r2	240 r2	330 r4	230 r3

■ 9)2974

◪ 6)1323

◪ 6)4862

◪ 3)812   ◪ 4)962   ■ 7)141   ◪ 3)1980

◪ 2)1521   ◪ 7)4836   ◩ 5)352   ◪ 8)964

◪ 5)953   ◪ 3)1322   ■ 4)923   ◪ 8)7523

45 DIVISION

701 r2	503 r8	808 r5	602 r1	607 r1	506 r1	607 r1	101 r4	409 r2	906 r1
606 r1	906 r1	608 r2	906 r1	608 r2	707 r1	402 r5	404 r2	101 r4	707 r1
409 r2	503 r8	808 r5	707 r1	506 r1	409 r2	606 r1	607 r1	701 r2	306 r2
101 r4	906 r1	402 r5	906 r1	402 r5	602 r1	409 r2	707 r1	101 r4	404 r2
503 r8	701 r2	606 r1	707 r1	506 r1	607 r1	101 r4	607 r1	608 r2	606 r1
808 r5	306 r2	506 r1	409 r2	404 r2	402 r5	707 r1	701 r2	409 r2	607 r1
707 r1	608 r2	404 r2	101 r4	602 r1	306 r2	101 r4	906 r1	808 r5	707 r1
409 r2	606 r1	607 r1	701 r2	607 r1	608 r2	503 r8	701 r2	404 r2	506 r1
602 r1	602 r1	404 r2	808 r5	707 r1	506 r1	409 r2	409 r2	306 r2	608 r2
306 r2	506 r1	906 r1	402 r5	306 r2	701 r2	602 r1	606 r1	707 r1	101 r4

■ $7\overline{)4243}$

◪ $8\overline{)6469}$

◪ $9\overline{)3638}$

◨ $4\overline{)3625}$ ◧ $7\overline{)4258}$ ■ $5\overline{)2047}$ ◪ $3\overline{)1519}$

◪ $9\overline{)4535}$ ◪ $5\overline{)3536}$ ◨ $4\overline{)1226}$ ◧ $7\overline{)4909}$

◨ $2\overline{)1215}$ ◨ $6\overline{)2417}$ ■ $3\overline{)1807}$ ◪ $8\overline{)812}$

46   DIVISION

11	2	12	14	9	3	6	16	4	5
4	5	13	15	8	9	7	10	11	2
3	6	16	12	10	13	16	4	14	3
5	9	8	3	4	2	3	9	8	11
12	15	7	5	9	8	13	10	7	14
11	2	4	14	8	9	12	2	6	15
14	8	9	3	13	10	3	8	9	12
3	13	10	7	2	6	5	11	10	3
7	16	6	14	8	9	4	16	4	5
12	5	11	15	3	9	13	15	7	16

■ $21\overline{)189}$

◣ $34\overline{)170}$

◩ $26\overline{)104}$

◤ $19\overline{)209}$ ◢ $99\overline{)198}$ ■ $35\overline{)105}$ ◪ $33\overline{)495}$

◣ $12\overline{)144}$ ◣ $44\overline{)264}$ ◺ $43\overline{)559}$ ◿ $11\overline{)176}$

◺ $25\overline{)175}$ ◿ $16\overline{)224}$ ■ $24\overline{)192}$ ◪ $82\overline{)820}$

21	28	47	81	21	28	82	24	21	28
33	19	73	28	19	37	33	42	48	33
21	34	37	21	34	81	21	72	81	21
33	28	42	48	33	28	19	73	28	33
28	21	72	92	47	24	34	48	21	28
33	47	73	19	92	42	48	82	24	33
19	81	72	81	28	33	34	37	42	37
82	37	34	24	21	28	72	92	19	92
47	92	33	42	48	47	73	21	34	48
81	28	21	19	81	34	37	33	28	82

■ $85\overline{)1785}$

◣ $75\overline{)2775}$

◥ $42\overline{)1764}$

◩ $36\overline{)1692}$ ◪ $32\overline{)2592}$ ■ $62\overline{)2046}$ ◥ $33\overline{)1584}$

◣ $17\overline{)1394}$ ◣ $82\overline{)2788}$ ◩ $27\overline{)1944}$ ◪ $23\overline{)1679}$

◩ $56\overline{)1064}$ ◪ $14\overline{)1288}$ ■ $42\overline{)1176}$ ◥ $65\overline{)1560}$

13	27	7 r4	12	301	233	50	13	27	7 r4
27	7 r4	3	22	3	50	55 r12	143	7 r4	13
13	131	536 r1	131	301	0	7 r5	55 r12	7 r5	27
7 r4	43	7 r4	22	13	27	233	7 r4	43	7 r4
12	22	27	12	7 r5	131	50	7 r4	0	143
0	50	13	55 r12	536 r1	55 r12	22	27	131	301
13	43	7 r4	143	27	7 r4	3	13	43	13
27	55 r12	143	233	50	12	22	12	301	27
7 r4	13	55 r12	7 r5	0	536 r1	3	22	7 r4	13
13	27	7 r4	233	50	131	536 r1	13	27	7 r4

■ $5\overline{)65}$

◪ $13\overline{)650}$

◣ $18\overline{)0}$

◸ $12\overline{)144}$ ◪ $13\overline{)3913}$ ■ $19\overline{)513}$ ◤ $5\overline{)715}$

◣ $3\overline{)699}$ ◢ $40\overline{)2212}$ ◹ $27\overline{)81}$ ◪ $3\overline{)1609}$

◸ $5\overline{)655}$ ◩ $21\overline{)462}$ ■ $5\overline{)39}$ ◤ $81\overline{)572}$

SOLUTION KEY

1 ADDITION

21
24
17

27	23	22	26
25	18	13	16
10	15	19	20

2 ADDITION

86
75
73

69	57	89	47
96	21	25	92
99	85	26	55

3 ADDITION

75
47
92

63	62	83	85
74	91	77	96
81	43	72	57

4 ADDITION

178
189
120

175	134	183	168
197	101	198	155
176	135	177	172

5 ADDITION

976
799
929

243	544	435	446
888	669	967	389
462	445	787	777

6 ADDITION

953
452
878

971	854	792	215
458	486	696	593
536	693	785	864

7 ADDITION

662
901
733

826	964	833	616
600	908	803	761
647	563	724	593

8 ADDITION

1465
1252
1522

1074	1377	1602	1546
1123	1258	1513	1371
1332	1930	1334	1776

9 ADDITION

152
189
175

200	209	215	188
173	212	219	207
158	149	186	139

10 ADDITION

2218
1987
2365

1252	1323	1015	1734
1622	1234	2774	1542
1336	1098	1232	1403

11 ADDITION

6435
8169
4772

9553	7697	3945	7004
8323	5694	4398	3761
8300	8425	3452	5736

12 REVIEW

16
36
142

112	123	19	77
1212	579	134	1009
122	1432	8	48

13 SUBTRACTION

2
11
16

7	3	4	14
10	12	8	5
1	9	6	15

14 SUBTRACTION

82
42
96

85	45	64	70
55	34	91	32
63	60	97	24

15 SUBTRACTION

24
11
20

61	33	12	10
36	14	43	13
26	8	17	2

16 SUBTRACTION

700
311
116

200	241	132	212
231	100	223	612
323	777	521	125

17 SUBTRACTION

59
46
37

69	26	88	53
19	28	76	56
77	18	48	84

18 SUBTRACTION

73
26
25

75	58	67	8
6	22	9	32
68	27	69	37

19 SUBTRACTION

389
927
516

247	734	128	628
667	417	189	464
722	239	444	369

20 SUBTRACTION

419
349
309

358	527	617	228
459	438	338	717
546	828	259	827

21 SUBTRACTION

53
452
862

63	91	162	783
493	675	880	195
795	232	791	86

22 SUBTRACTION

683
677
257

286	366	179	488
277	686	587	585
588	369	386	339

23 SUBTRACTION

769
38
335

239	279	468	568
137	174	39	368
166	215	638	466

24 REVIEW

28
427
16

77	231	8	185
25	67	15	567
761	477	22	84

25 MULTIPLICATION

25

9

3

54	16	45	81
72	36	63	30
35	64	0	12

26 MULTIPLICATION

213

637

180

0	86	159	96
49	129	148	189
120	288	168	426

27 MULTIPLICATION

168

295

344

416	256	335	462
372	180	456	136
324	350	252	352

28 MULTIPLICATION

1278

1878

436

657	6496	942	858
987	1578	1256	972
460	672	1652	1251

29 MULTIPLICATION

2114

1212

918

6318	3521	728	5472
1218	410	1025	520
7218	327	624	1421

30 MULTIPLICATION

306

2346

768

1056	1155	2286	504
1928	2328	1768	1720
1740	2380	5220	1449

31 MULTIPLICATION

2352

1368

2170

3356	1338	1645	6265
9464	1395	2844	1824
7950	1692	10,936	1167

32 MULTIPLICATION

253

308

1749

483	1606	989	2046
852	736	714	1008
682	528	756	837

33 MULTIPLICATION

900

4500

2000

2500	400	1600	100
800	6300	1000	4200
3600	300	1200	1500

34 MULTIPLICATION

4187

2184

6256

2345	2340	2210	2646
2106	6045	2052	1566
1125	2250	4928	2652

35 MULTIPLICATION

13,685

37,668

23,460

33,885	49,082	19,980	11,418
18,496	15,330	36,426	34,224
23,532	24,360	14,196	30,618

36 REVIEW

810

850

9200

30	64	72	42
280	100	636	306
330	124	56	1020

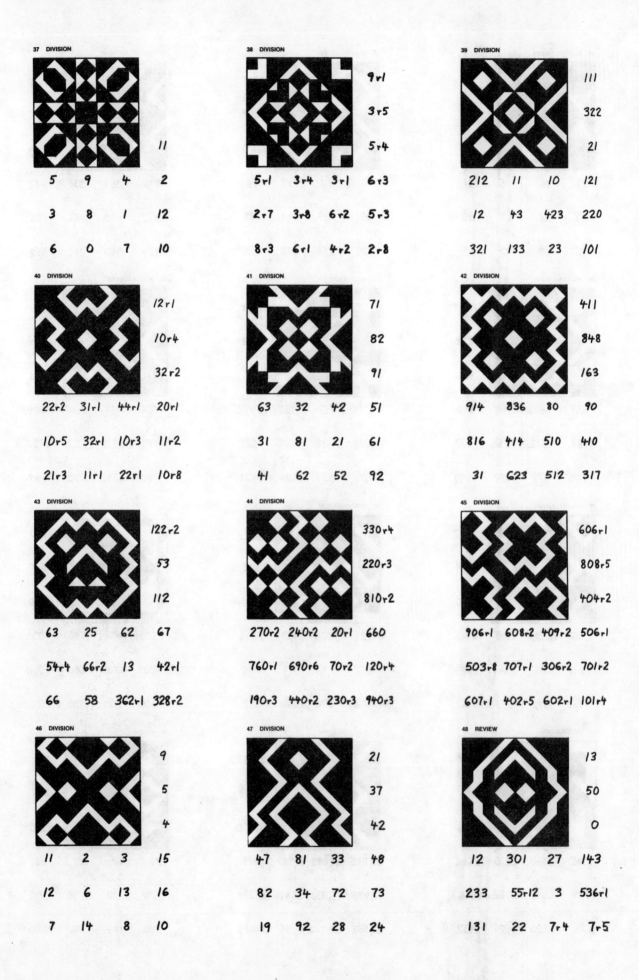

37 DIVISION

11

5	9	4	2
3	8	1	12
6	0	7	10

38 DIVISION

9r1

3r5

5r4

5r1	3r4	3r1	6r3
2r7	3r8	6r2	5r3
8r3	6r1	4r2	2r8

39 DIVISION

111

322

21

212	11	10	121
12	43	423	220
321	133	23	101

40 DIVISION

12r1

10r4

32r2

22r2	31r1	44r1	20r1
10r5	32r1	10r3	11r2
21r3	11r1	22r1	10r8

41 DIVISION

71

82

91

63	32	42	51
31	81	21	61
41	62	52	92

42 DIVISION

411

848

163

914	836	80	90
816	414	510	410
31	623	512	317

43 DIVISION

122r2

53

112

63	25	62	67
54r4	66r2	13	42r1
66	58	362r1	328r2

44 DIVISION

330r4

220r3

810r2

270r2	240r2	20r1	660
760r1	690r6	70r2	120r4
190r3	440r2	230r3	940r3

45 DIVISION

606r1

808r5

404r2

906r1	608r2	409r2	506r1
503r8	707r1	306r2	701r2
607r1	402r5	602r1	101r4

46 DIVISION

9

5

4

11	2	3	15
12	6	13	16
7	14	8	10

47 DIVISION

21

37

42

47	81	33	48
82	34	72	73
19	92	28	24

48 REVIEW

13

50

0

12	301	27	143
233	55r12	3	536r1
131	22	7r4	7r5